I0817014

Sharks

Nurse Sharks

by Julie Murray

Dash!
LEVELED READERS
An Imprint of Abdo Zoom • abdobooks.com

Level 1 – Beginning
Short and simple sentences with familiar words or patterns for children who are beginning to understand how letters and sounds go together.

Level 2 – Emerging
Longer words and sentences with more complex language patterns for readers who are practicing common words and letter sounds.

Level 3 – Transitional
More developed language and vocabulary for readers who are becoming more independent.

THIS BOOK CONTAINS RECYCLED MATERIALS

abdobooks.com

Published by Abdo Zoom, a division of ABDO, PO Box 398166, Minneapolis, Minnesota 55439.

Printed in the United States of America, North Mankato, Minnesota.
102019
012020

Photo Credits: Alamy, iStock, Minden Pictures, Seapics.com, Shutterstock
Production Contributors: Kenny Abdo, Jennie Forsberg, Grace Hansen, John Hansen
Design Contributors: Dorothy Toth, Neil Klinepier, Victoria Bates

Library of Congress Control Number: 2019941289

Publisher's Cataloging in Publication Data

Names: Murray, Julie, author.
Title: Nurse sharks / by Julie Murray
Description: Minneapolis, Minnesota : Abdo Zoom, 2020 | Series: Sharks | Includes online resources and index.
Identifiers: ISBN 9781532129223 (lib. bdg.) | ISBN 9781098220204 (ebook) | ISBN 9781098220693 (Read-to-Me ebook)
Subjects: LCSH: Nurse shark--Juvenile literature. | Sharks--Juvenile literature. | Fish--Juvenile literature. | Ocean animals--Juvenile literature. | Top predators--Juvenile literature. | Carnivores--Juvenile literature.
Classification: DDC 597.33--dc23

Table of Contents

Nurse Sharks

Nurse sharks live in warm waters. They are found around the world.

They have **thick** bodies. Some grow 14 feet (4.3 m) long.

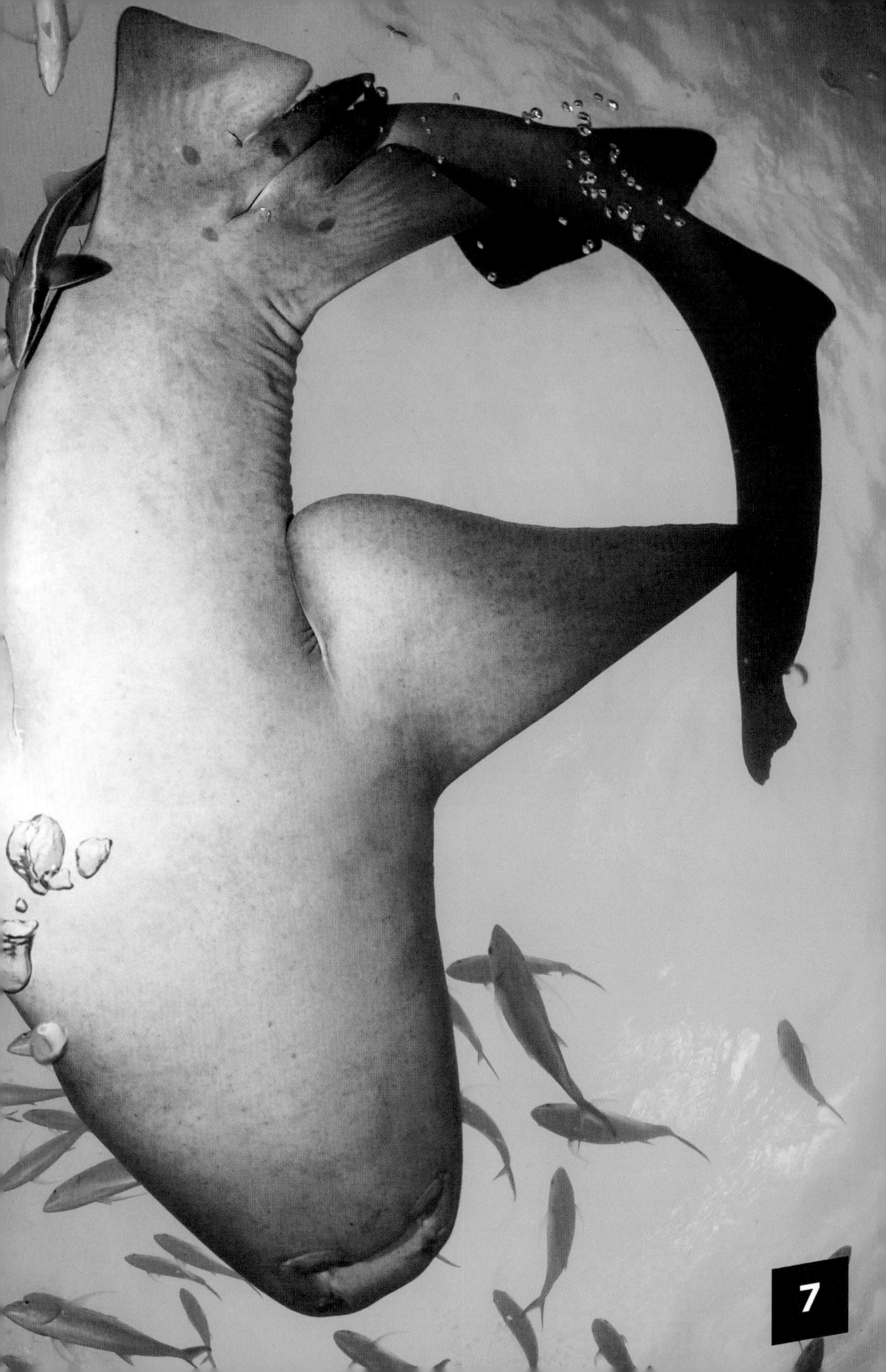

They are tan to dark brown in color.

Their skin has tooth-like scales. They are called **denticles**. These protect their skin.

During the day, they rest in groups. Up to 40 of them will pile together.

They are active at night. They swim slowly on the ocean floor.

They have whisker-like organs called **barbels**. These help them find food.

They suck up their food like a vacuum cleaner. They eat clams and shrimp. They also eat squid.

They are calm sharks. But they will attack if they feel **threatened**.

More Facts

- Nurse sharks do not travel far. They stay in a small area for their entire lives.
- Nurse sharks are calm and slow moving. This is why they are often called the laziest sharks.
- It can look like they are walking on the sea floor. They use their fins like feet.

Glossary

barbel – a slender, whisker-like organ on the heads of certain fishes.

denticle – a small tooth-shaped scale that is found on certain fishes, like sharks and rays.

thick – large from one end to the other.

threatened – caused to feel at risk or in danger.

Index

Online Resources

To learn more about nurse sharks, please visit **abdobooklinks.com** or scan this QR code. These links are routinely monitored and updated to provide the most current information available.